Omnimics:

An Executive Playbook for Artificial Intelligence and Analytics

J. Tod Fetherling
Author/Publisher
7011 Ellendale Drive
Brentwood, TN. 37027

ISBN - 979-8-9928531-8-6

April 16, 2025

OMNIMICS

An Executive Playbook
for
Artificial Intelligence
and Analytics

By J. Tod Fetherling

TABLE OF CONTENTS

INTRODUCTION

Omnimics is the state of knowing the answer before you ask a question.

We live in an exciting time. We can now understand complex issues using a voice command or a "prompt." This technology provides us with answers if we ask the right questions. The new world we inhabit brings new challenges. These challenges parallel past technological innovations like the World Wide Web (WWW), email, video, and broadband. Each of these technologies created a gap between individuals and organizations with access and those without.

We are now amid the next evolution of analytics: AI-Infused Analytics. In this realm, we should be able to anticipate and devise solutions even before individuals and organizations pose their questions.

About a decade ago, the term Omnimics came to me after a meeting with a hospital Chief Executive Officer (CEO). He knew what he wanted and had plenty of information from various, albeit conflicting, software systems. However, he still didn't receive the insights needed to run the hospital. Sound familiar? Too much information and not enough intelligence. He wanted to know everything (OMNI), and even though he had just met me, he expected me to answer his questions on the fly (Mini-Analytics). So, I combined the two concepts and coined the term Omnimics – knowing everything before you even ask the question.

I first used the term eight years ago in my TNHIMSS/ Belmont HIT Accelerator Strategy and Analytics Class. Over the past eight years, many students have engaged in this class and discussion with open minds to envision what the future might look like in a digital age where intelligence is "on demand."

This book is a game plan for any executive looking to successfully implement an Omnimics solution.

To successfully execute this playbook, you must understand the concept, ensure the necessary infrastructure is in place, have the correct data feeds, and articulate your desired outcome. Like so many things in business today, the technology is already there, waiting for someone to present the problems to be solved.

BASIC REQUIREMENTS

Many aspects of an Omnimics strategy can leverage existing investments in Data, Infrastructure, Security, and Governance. We will briefly examine each area of technology, people, and processes. Let's begin with infrastructure.

Infrastructure Requirements

Modern architecture is essential for successfully implementing Omnimics strategies. Traditional architectures tend to be rigid and often struggle to manage dynamic data flows from multiple sources, where incoming data initiates real-time processing for advancements in data and models.
An entire book could be dedicated to designing an effective Omnimics infrastructure. However, in this book, we will focus on the key requirements needed to achieve the desired outcomes.

Numerous cloud-based infrastructure options exist, including Amazon Web Services (AWS), Microsoft Azure, Google Cloud, and IBM, along with hundreds of data vendors like Snowflake and Databricks. Many companies focus on assisting organizations in designing and implementing scalable infrastructure solutions tailored to their needs.

Beyond the Server

The conversation around infrastructure extends beyond the server. Recently, all discussions have centered on the processor (Nvidia, ARM, etc.). A single weak link—whether in processing power, server configuration, cloud vs. on-prem deployment, application architecture, data management, security, or documentation—can lead to system failures, whether gradual or sudden.

Infrastructure in technology is akin to the foundation of your house. If this foundation fails, the house collapses, sometimes slowly over the years or suddenly during a storm. The processor, server, configuration, cloud/on-premises, application, data, security, and documentation must all align toward the same goal – the result. Advanced analytics is impossible without a comprehensive understanding of how these elements fit together.

Below is a diagram referenced to assist technology teams in constructing a modern, flexible, and robust architecture.

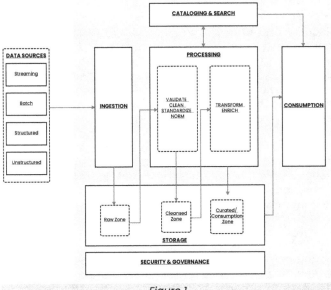

Figure 1

Cataloging and Security

Once you have your infrastructure in place, the next area of concern is cataloging and searching data assets. This step is essential yet often overlooked. Creating and maintaining a catalog of your assets while ensuring they are searchable enhances the overall usefulness of the plan for everyone in the organization who needs to learn and grow from your work. This accessibility benefits all those you choose to grant access. It is important to create metadata (what) when you begin to build the data assets and feeds for an Omnimics solution.

Security and Data Governance

Security and data governance should be a top priority, not an afterthought. If you prioritize this as one of your initial steps rather than the final one (as most people do), you'll save yourself hundreds, if not thousands, of hours in meetings. Building consensus on who should have access to specific assets and which processes are involved is crucial for streamlining decision-making processes.

I've observed well-intentioned teams repeatedly struggle with governance issues. Why? Because getting governance right is challenging and demands patience. Often, governance challenges arise from insufficient documentation within an organization.

The Role of Documentation

Documentation enables the organization to see and understand what data is generated or imported, allowing all users to observe how the data is transformed more transparently. At this stage, it is essential to ensure that every data set includes a data dictionary, a key tool for enhancing consistency and usability across teams.

The CISO and Beyond

To ensure that security is enabled and activated, engage the Chief Information Security Officer (CISO) early and often in the process. Depending on government regulations in your industry, you may need to involve auditing and legal teams to determine the necessary security levels for this solution. With a fully integrated Omnimics solution, you'll have access to any answer even before you ask the question. This could include sensitive information, such as internal memos and emails. It is important to carefully consider what you do and do not want included in the final solution.

Data: The Core of the Game Plan

Without data, you're already falling behind the industry leaders. If you have ample data but lack insights, the Omnimics solution can help. It is crucial that all key data sources integrate with the platform. I won't elaborate on each data source listed below; instead, I will provide a list of potential data sources that I have found useful in the past and that you may wish to consider for inclusion in the final solution.

Markets

> Market Trends
> Economic Shifts
> Consumer Behavior

Emerging Technologies

Government/Regulatory/Political Landscape

Competition

> Strategies
> Market Positioning
> Strengths and Weaknesses
> Pricing

Customer Data

> Demographics
> Psychographics
> Customer Preferences
> Customer Acquisition Costs
> Pipeline/Conversion

Product/Service Portfolio

> Product Market Fit
> User Experience

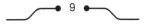

Business Plans

Financials / Performance

>   Income Statement
>   Balance Sheet

Operations

Historical Industry Data

Defining the Right Questions

Omnimics focuses on understanding what you want answered. This may seem like the simplest question, but it often trips up many executives when they seek to implement a strategy around Omnimics. To say you want everything is not valid in this world. I approach the solution as one with finite resources (Time and Money). As such, leaders must make some initial tradeoffs as they build this solution.

Understanding the "Why" behind the data is critical. The CEO's priorities may differ completely from the Chief Financial Officer (CFO), Chief Operations Officer (COO), Chief Revenue Officer (CRO), Chief Product Officer (CPO), or Chief Marketing Officer (CMO). Have these leaders' established alignment on key metrics? Are managers and employees on the same page?

From an organizational standpoint, each leader has unique priorities. These must be understood and ranked accordingly to integrate them effectively into the Omnimics solution.

EXECUTING AN OMNIMICS STRATEGY

What is Artificial Intelligence to you, and how should you utilize these technologies to effectively and efficiently position your company for success? This is always the first question I start with because many people define AI in various ways. It has become a catch-all for emerging technologies. Although it is one of the most exciting advancements since email and web browsers, it is not the only emerging technology that executive leaders should be informed about. Does it impact almost all aspects of information technology? The answer is yes.

What are your Company's Vision and Values?

For me, this is the most important question in this space. If you truly understand why you are pursuing this business in the first place, applying the Omnimics Solution allows you to accelerate the achievement of your vision. Values play a critical role in Omnimics as they provide the guardrails for an organization utilizing this technology.

Organizational Structure

Where should AI reside within your company? Many assume it belongs in the IT department, but successful AI initiatives can be found in Data & Analytics, Strategy, Operations, or even as an independent function.

Establish a clear team hierarchy, define roles and responsibilities for each team member, set communication protocols (frequency), and implement a comprehensive decision-making framework. Since AI initiatives often require significant budget allocations for both operating and capital expenses, employing a structured approach is critical.

Standard Operating Procedures (SOPs).

Governance is critical. An entire game plan should be constructed for Governance alone. Many companies struggle with this, just as they do with documentation—everyone acknowledges its importance, but few prioritize it properly.

Depending on the sensitivity of the data you are dealing with, ensure you have a crisis management plan. It is also important to conduct a Security Risk Analysis. You should perform four types of security audits. It is critical to include your organization's CISO in the planning and execution of any security audits.

There are four main types of risk assessments that organizations utilize:

- Qualitative
- Quantitative
- Subjective
- Objective

Leadership Strategies

A new role emerges: the Chief AI Officer (CAIO) or Chief Analytics Officer (CAO). As AI-driven strategies evolve, a well-rounded team should include:

- **A Technologist** – Someone with hands-on expertise in building AI solutions.
- **A Product Manager/Owner** – Someone with experience in developing tech products and project management.

- **An Ethics Officer (Humanist)** – A professional who can critically evaluate the impact of AI from a societal and ethical standpoint.
- **An Analyst** – Business and financial experts who can assess models, track KPIs/OKRs, and translate data into meaningful insights.
- **A Designer/Creative** – A user experience (UX) expert with knowledge of React frameworks and AI interaction design.
- **A Leader** – This could be one of the above roles or a dedicated AI project lead with a strong background in product development and business strategy.

People and organization

I recommend working in two-week sprints. This gives the team enough time to produce work and still have some time for curiosity. Communication Touchpoints are critical for the team and the organization. The ideal cadence is based on your team and what the organization expects out of an R&D type of department.

Business Plans and Goals

AI needs a separate product and data roadmap. Make sure to include SMART Goals for the organization and tie the results to the overall strategy for the organization.

Performance Indicators and Metrics (Key Performance Indicators (KPIs), Objectives and Key Results (OKRs), etc.)

I usually start with profitability and customer adoption. Suppose your goal of implementing Omnimics is not to improve the business, enhance customer experience, or

boost profitability. In that case, you may need to reconsider whether you are doing this for the right reasons in the first place.

Systems, Tools, and Techniques

This is likely to generate the most discussion within the organization. Should you leverage existing infrastructure or move forward and build on a modern architecture? If you choose the former, I believe you will regret it when you need more capacity, and the current infrastructure cannot support you. This is also connected with budgets (see above).

The tools are booming. I would do a disservice by trying to cover all the new tools in this space because they are being released every day. If it's not a new tool, it's a version upgrade.

Document, Document, Document. I can say it enough. "When you wink at a person in the dark, only you know what you are doing." Creating documentation while you are building gives you an extra opportunity to share these documents with other departments and senior management/Board of Directors. Even if people don't fully read it, it will be useful for all who come after you.

As you embark on this journey, it's important to understand and stay updated with the new tools as they are released. Even better, consider partnering with these companies and participating in beta programs to preview the new products before they go into production. Below, I've used the same prompt on multiple platforms.

Here are some prompts and AI Engines that should help you on your journey.

What tools do you recommend to help me stay up to date with AI

To stay current with AI developments, leverage a combination of curated tools and strategies:

Core AI Assistants & Search Tools
– **Perplexity**: AI-powered search engine for real-time answers and research aggregation [2] [7⊕] .
– **Claude/Gemini/ChatGPT**: Leading LLMs for synthesizing information and exploring technical concepts [4] [7⊕] .
– **Google Scholar Alerts**: Track academic papers and breakthroughs [1] [3] .

Content Aggregation & News
– **Newsletters**: Subscribe to specialized AI newsletters (e.g., Armand Ruiz's curated updates) [6] .
– **Twitter/X**: Follow experts like Andrew Ng, Yann LeCun, and organizations like OpenAI/DeepMind [1] [3] .
– **Papers with Code**: Discover trending research papers with implementation code [1] .

Knowledge Management
– **Notion AI Q&A**: Internal knowledge base tool for instant answers from organizational data [2] .
– **Guru**: AI-powered platform for centralized, verified company knowledge [2] .
– **Gamma**: AI presentation tool for summarizing trends into shareable formats [7⊕] .

Automation & Productivity
– **Zapier Chatbots**: Build custom AI assistants for workflow automation [3] .
– **Fireflies.ai/Otter.ai**: Automate meeting notes and extract actionable insights [7⊕] .
– **Cursor**: AI coding assistant for staying current with development practices [7⊕] .

Community & Learning
– **Meetup/LinkedIn Groups**: Join AI-focused communities for event updates [5] [9] .
– **Conferences**: Attend NeurIPS, ICML, or local AI meetups [5] .
– **AI Competitions**: Participate in Kaggle challenges to test cutting-edge techniques [5] .

For efficiency, dedicate 1-2 hours weekly to reviewing prioritized resources [1] [9] , and use tools like **NotebookLM** for interactive research summarization [7⊕] .

What tools do you recommend to help me stay up to date with AI

There are several effective tools to help you stay up to date with AI developments. Here's what I recommend:

1. **AI newsletters** - Subscribe to publications like The Batch by DeepLearning.AI, Import AI by Jack Clark, or The Algorithm by MIT Technology Review for curated updates.

2. **AI research platforms** - Follow arXiv.org (particularly the cs.AI, cs.LG, and cs.CL categories) and Papers With Code to track the latest research papers and implementations.

3. **Industry blogs** - Major AI companies like Anthropic, OpenAI, Google AI, and Meta AI publish insightful blog posts about their research and products.

4. **Podcasts** - Listen to shows like Lex Fridman Podcast, TWIML AI Podcast, or The AI Alignment Podcast for in-depth discussions with researchers.

5. **Social media** - Follow AI researchers, practitioners, and organizations on Twitter/X or LinkedIn. Look for hashtags like #AIresearch or #MachineLearning.

6. **GitHub repositories** - Star and watch popular AI repositories to see new techniques and implementations in real-time.

7. **Conference proceedings** - Browse papers from major AI conferences like NeurIPS, ICML, ICLR, and ACL.

8. **AI news aggregators** - Sites like Hacker News (with AI filters) or specialized platforms like AI Trends compile relevant stories.

Claude 3.7 Sonnet

Staying current with AI can feel like trying to drink from a firehose—things move fast! But with the right tools, you can turn that firehose into a manageable stream. Here's a mix of tools, feeds, and strategies to keep you sharp:

📰 Newsletters (low effort, high signal)

These deliver curated updates to your inbox:

- **The Batch (by DeepLearning.AI)** – Weekly, high-quality summaries of AI research, products, and trends.
- **Import AI (by Jack Clark)** – Focused and thoughtful takes on AI developments.
- **TLDR AI** – Short, daily digests on AI news, tools, and papers.
- **Ben's Bites** – Great daily AI newsletter with humor and a strong product/tools focus.

🔬 Research Trackers & Summaries

Stay updated on papers and breakthroughs:

- **Papers with Code** - New research papers + code, sortable by trending topics or fields.
- **arXiv Sanity** – Curated arXiv feed with filters and recommendations (by Andrej Karpathy).
- **Explainpaper** – Helps you understand dense research papers using natural language explanations.

🛠️ Tools & Aggregators

Perfect for exploring new AI products and experiments:

- **There's an AI for That** – Search engine for AI tools by category or use case.
- **FutureTools.io** – Curated AI tools and news, with filters and "tool of the day."
- **Supertools** – Discover powerful AI tools by workflow (e.g., coding, writing, design).

📱 Social Media & Forums

For real-time updates and expert takes:

- **Twitter/X** – Follow researchers (e.g., @karpathy, @sama, @ylecun, @quocle) and devs.
- **Reddit** – Subreddits like r/MachineLearning, r/ArtificialIntelligence, and r/LocalLLaMA.
- **Hacker News** – Daily AI-related threads and discussion.

📺 YouTube & Podcasts

For digestible, deep-dives and discussions:

- **Two Minute Papers (YouTube)** – Fun, quick summaries of cutting-edge research.
- **Lex Fridman Podcast** – Long-form, in-depth interviews with top AI minds.
- **Practical AI Podcast** – Covers tools, news, and real-world AI applications.

💡 Bonus Tip: Make a Weekly Routine

Set aside a block (30–60 mins) each week to:

- Skim newsletters.
- Save papers from Papers with Code.
- Watch a short video or listen to a podcast.

Applying the Vairos Strategy Process™ as a Strategic Framework

Using the Vairos Strategy Process ™ as a Strategic Framework, one can develop a thorough understanding of AI capabilities (what they can do now and what they can do in the next few years) by gathering the required information from within the organization and structuring these ideas into a process by which the operators can implement a logical plan which can be followed by management, customers, investors, and the employees.

Phase 1: Assess

Step 1: Inventory Current and Planned Investments

Begin by evaluating your organization's existing and anticipated investments in AI. This assessment should cover:

- Architecture
- Infrastructure
- Data
- Security
- Governance

AI INITIATIVES

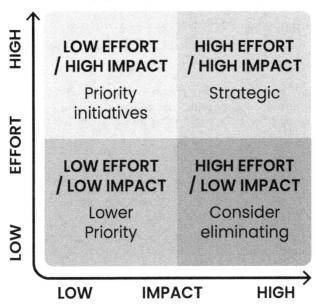

Step 2: Rank Effort vs. Impact

Use a 2x2 matrix to categorize AI initiatives based on effort and impact:

- Quadrant 1: Low Effort / High Impact (Priority initiatives)
- Quadrant 2: High Effort / High Impact (Strategic/ Resource-intensive)
- Quadrant 3: Low Effort / Low Impact (Lower priority)
- Quadrant 4: High Effort / Low Impact (Consider eliminating)

Conduct this ranking exercise progressively:

1. Start individually (as the leader).
2. Engage direct reports to assess alignment.
3. Expand to managers for broader input.
4. Include employees to capture operational insights.

This structured approach ensures organizational alignment on AI priorities. Depending on your company's size, consider outsourcing this moderation to a consulting firm.

Step 3: Conduct Research

Gather intelligence through multiple channels:

- Engage vendors (e.g., OpenAI, Google, Microsoft, IBM, Palantir, Nvidia).
- Interview employees to understand internal AI readiness.
- Host an internal hackathon to explore AI applications.
- Leverage AI tools to summarize findings and generate initial recommendations.

Step 4: Benchmark Against Industry Peers

Compare your organization's AI investments and strategies with competitors. Develop a Spend and Impact Analysis (SIA) to evaluate ROI and decision-making effectiveness.

Phase 2: Decide

Step 5: Make Strategic Decisions

Based on your research and benchmarking, finalize your AI strategy:

- Define key objectives and desired outcomes.
- Set a timeline for implementation.
- Allocate a budget to support execution

Phase 3: Execute

Step 6: Implement with Focus and Accountability

Execution requires:

- A clear ownership structure to drive accountability.
- Rigorous tracking and reporting on progress
- A culture of iteration, ensuring continuous optimization of AI initiatives.

ANALYTICS

To provide context, let's review the Analytics Progression and how we arrived at this point. Over 80% of the work analytics professionals perform daily falls under Descriptive Analytics—from running cross-tabulations to creating dashboards for executives. A significant portion of their time is spent preparing data and delivering key insights for decision-making.

As we move from Descriptive to Diagnostic, the basics of Omnimics start to form the basis of the solution. We start to ask why something is happening.

The next step requires a higher level of talent and skill. Asking the right questions to create predictive analytics is essential for success and budget management. In this phase, there is a potential to spend significantly more on processing and computing time. I always encourage individuals to start small with a known outcome. This method enables you to answer the initial question confidently, understand the costs involved in obtaining an answer, and identify the people, processes, and technology available for this type of work.

The last phase is prescriptive. This approach only works if you can intervene in the process and guide the user or customer toward a new, improved state- helping them save money, helping them make money, and helping them make the right choices. The closer you can position prescriptive analytics to the point of sale, such as Amazon's recommendations during checkout, the better the outcomes. This is particularly true in healthcare, where many of these decisions will affect the quality of life for years to come.

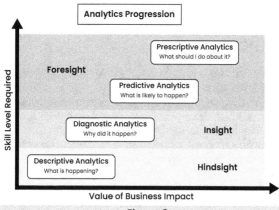

Figure 2

- **Descriptive analytics:** This type determines what is happening based on existing data.

- **Diagnostic analytics:** This type goes one step further to determine why a specific situation happened.

- **Predictive analytics:** This type looks across a broader set of data, perhaps over a longer period, to see trends and examples and then uses that historical information to predict future occurrences.

- **Prescriptive analytics:** This type goes beyond prediction to suggest how to best change future situations to meet your goals.

Insights

I was introduced to DIKW: Data, Information, Knowledge, Wisdom at StrataRX in San Francisco in 2012. Some attribute this concept originally to T.S. Eliot's "The Rock. " Others credit Russell Ackoff, a systems theorist, who formalized and presented the DIKW (Data-Information-Knowledge-Wisdom) hierarchy in his 1989 article "From Data to Wisdom."

I was floored by how simple this concept is and how powerfully it explains the progression through the four concepts.

The DIKW pyramid represents the progression from raw data to actionable wisdom:

AI INITIATIVES

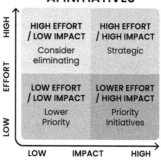

Figure 3

- **Data:** Raw, unprocessed facts and figures without context.

- **Information:** Organized and contextualized data that answers basic questions like who, what, where, and when.

- **Knowledge:** Analyzed and interpreted information revealing patterns, trends, and relationships.

- **Wisdom:** The ability to make well-informed decisions based on understanding the underlying knowledge.

This hierarchy illustrates how each level builds upon the previous one, adding value and meaning to the initial data. For example, in fitness tracking:

- Data: Raw step count, heart rate, and sleep duration

- Information: Daily step count, average heart rate, and hours of sleep per night

- Knowledge: Patterns linking increased step count to improved sleep quality

- Wisdom: Making informed decisions about exercise routines and lifestyle changes

Asking the Right Questions = Gain Understanding and Wisdom

The ability to ask thoughtful questions is often considered a key aspect of wisdom. This concept emphasizes that:

- Wisdom lies in formulating the right questions rather than having all the answers.
- Asking intentional questions can lead to deeper understanding and insights.
- The highest expression of wisdom may be the perfectly formed question.

Dr. Lance Hawley suggests that cultivating wisdom involves asking intentional, thought-provoking questions to shape our thinking and way of life. This approach aligns with the Socratic method and the classical education goal of teaching students to think critically (footnote).

Going from Qualitative to Quantitative

The process of converting qualitative data into a quantitative form is known as quantizing. This transformation is often used in mixed methods research to:

- Facilitate pattern recognition in qualitative data.
- Extract meaning and account for all data.
- Verify interpretations and document analytic moves.

However, quantizing involves several considerations:

- Deciding what and how to count.
- Balancing numerical precision with narrative complexity.
- Recognizing that counting is not always an unambiguous process.

The value of quantizing can be understood through perspectives such as "conditional complementarity," "critical remediation," and "analytic alternation". These standpoints help clarify the benefits of converting qualitative data into quantitative form in research contexts.

Today's GenAI exemplifies qualitative responses from AI. The software is asked to provide a high-quality answer: What steps should I take to reboot my Mac?

Going from Qualitative to Quantitative, what if I could ask a search box to give the probability that I will be diagnosed with diabetes on a Tuesday and then have an emergency room visit in less than 7 days? In this example, we are looking for a numeric response, such as you are 3.5% likely to be diagnosed with diabetes on Tuesday and have an emergency room visit in less than 7 days.

The Future of AI-Driven Analytics

As AI evolves, we must recognize:

- We don't know what we don't know.
- Experimentation and iteration are key.
- A structured approach to analytics will drive better outcomes.

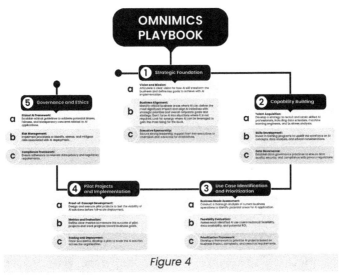

Figure 4

OMNIMICS PLAYBOOK

Key components for your Playbook:

Strategic Foundation:

- **Vision and Mission:**
  Articulate a clear vision for how AI will transform the business and define key goals to achieve with AI implementation.

- **Business Alignment:**
  Identify critical business areas where AI can deliver the most significant impact and align AI initiatives with strategic priorities and overall corporate goals and strategy. Don't force AI into situations where it is not required. Look for synergy where AI can be leveraged to gain the most bang for the buck.

- **Executive Sponsorship:**
  Secure strong leadership support from key executives to champion and advocate for AI initiatives.

Capability Building:

- **Talent Acquisition:**
  Develop a strategy to recruit and retain skilled AI professionals, including data scientists, machine learning engineers, and business analysts.

- **Skills Development:**
  Invest in training programs to upskill the workforce on AI concepts, data analysis, and ethical considerations.

- **Data Governance:**
  Establish data governance practices to ensure data quality, security, and compliance with privacy regulations.

Use Case Identification and Prioritization:

- **Business Needs Assessment:**
  Conduct a thorough analysis of current business operations to identify potential areas for AI application.

- **Feasibility Evaluation:**
  Assess each identified AI use case's technical feasibility, data availability, and potential ROI.

- **Prioritization Framework:**
  Develop a framework to prioritize AI projects based on business impact, complexity, and resource requirements.

Pilot Projects and Implementation:

- **Proof-of-Concept Development:**
  Design and execute pilot projects to test the viability of AI solutions before full-scale deployment.

- **Metrics and Evaluation:**
  Define clear metrics to measure the success of pilot projects and track progress toward business goals.

- **Scaling and Deployment:**
  Once successful, develop a plan to scale the AI solution across the organization.

Governance and Ethics:

- **Ethical AI Framework:**
  Establish ethical guidelines to address potential biases, fairness, and transparency concerns related to AI applications.

- **Risk Management:**
  Implement processes to identify, assess, and mitigate risks associated with AI deployment.

- **Compliance Framework:**
  Ensure adherence to relevant data privacy and regulatory requirements.

Example Use Cases for an AI Executive Playbook:

- **Customer Service:**
  Implementing AI-powered chatbots to improve customer support efficiency.

- **Sales Forecasting:**
  Utilizing AI to predict sales trends and optimize inventory management.

- **Fraud Detection:**
  Applying machine learning algorithms to identify real-time fraudulent transactions.

- **Personalized Marketing:**
  Leveraging AI to deliver targeted marketing campaigns based on customer data.

- **Predictive Maintenance:**
  Using AI to analyze sensor data and predict equipment failures before they occur.

Important Considerations that are unique to each organization:

- **Continuous Learning:**
  Regularly update knowledge on emerging AI technologies and best practices to stay ahead of the curve.

- **Communication and Change Management:**
  Effectively communicate AI initiatives to all levels of the organization to foster adoption and address concerns.

- **Collaboration:**
  Foster cross-functional teams to ensure successful AI implementation across different business units.

ARTIFICIAL INTELLIGENCE

The world of artificial intelligence expands daily, thanks to ongoing innovations that continually reshape how we interact with technology and each other. At its core, AI is a field driven by relentless creativity and scientific inquiry, where each breakthrough builds on prior advances to create systems that are more intuitive, powerful, and responsive. This iterative process not only advances individual technologies but also inspires cross-disciplinary pivots, leading to the emergence of entirely new applications and fields.

GENAI – Generative Artificial Intelligence

- Large Language Models (ChatGPT, Bard, LLaMA)
- Text to Image Models (DALL-E, Midjourney)
- Text to Video (SORA)

Predictive AI

Machine Learning

- Supervised
- Unsupervised
- Reinforcement Learning

Neural Networks and Deep Learning

Generative Adversarial Networks (GANS)

Computer Vision

Natural Language Processing (NLP)

The Landscape of Artificial Intelligence

Artificial Intelligence (AI) has revolutionized various aspects of our lives, from how we interact with technology to how we solve complex problems. This section explores different types of AI, each with its unique capabilities and applications.

Generative Artificial Intelligence (GenAI)

Generative AI is a subset of artificial intelligence that creates new content based on patterns learned from existing data. It can produce text, images, videos, or other forms of data, often in response to natural language prompts. GenAI has applications across numerous industries, including software development, healthcare, finance, and entertainment.

Large Language Models (LLMs)

Large Language Models are a type of GenAI specifically designed to understand and generate human-like text. Notable examples include:

- **ChatGPT:** Developed by OpenAI, ChatGPT excels in generating human-like text, language translation, and creative content creation.

- **Bard:** Google AI's creation, Bard, showcases versatility in handling complex queries and various text-based tasks.

- **LLaMA:** This model demonstrates exceptional critical thinking abilities and proficiency in programming-related tasks.

Text-to-Image Models

These AI models generate images from textual descriptions:

- **DALL-E:** Created by OpenAI, DALL-E produces realistic and clear images with high fidelity to the given prompts.
- **Midjourney:** Known for its artistic style, Midjourney generates images that often resemble digital paintings or illustrations.

Text-to-Video Models

- **SORA:** Developed by OpenAI, Sora is a text-to-video generator that creates realistic videos up to 1080p resolution from text prompts. It offers features like frame-by-frame storyboard control and the ability to remix assets.

Predictive AI

Predictive AI uses historical data to forecast future outcomes or behaviors. It's commonly used in finance, weather forecasting, and recommendation systems.

Machine Learning

Machine Learning is a subset of AI that enables systems to learn and improve from experience without being explicitly programmed. It includes several approaches:

- **Supervised Learning:** The algorithm learns from labeled data to make predictions or decisions.
- **Unsupervised Learning:** The system identifies patterns in unlabeled data.

Reinforcement Learning:

The AI learns through trial and error, receiving rewards or penalties for its actions.

Neural Networks and Deep Learning

Neural networks are computing systems inspired by biological neural networks. Deep Learning uses multiple layers of these networks to process complex data and make decisions.

Generative Adversarial Networks (GANs)

GANs consist of two neural networks competing against each other, typically used to generate highly realistic synthetic data.

Computer Vision

This field of AI focuses on enabling machines to interpret and understand visual information from the world, much like human vision. While facial recognition is often the most debated aspect within this category, the application of computer vision in agriculture, especially using drones, is quite impressive. Crop yields are likely to improve with technologies like Omnimics. We may be moving toward a future where we can comprehend the demand for various crops and plant only the necessary amount of food for the country each year. Naturally, we need an acceptable variance for farmers and merchants.

Natural Language Processing (NLP)

NLP allows machines to understand, interpret, and generate human language, enabling applications like machine translation, sentiment analysis, and chatbots. NLP is particularly useful in healthcare where there are still paper charts and extensive clinician notes being inputted by hand or voice/transcription. Another great use of NLP is within the survey world, where you need large blocks to be categorized.

Once NLP is deployed, creating sentiment analysis becomes easier. In my work with AWS Comprehend, the innovations that utilize this tool are significant and likely to be embedded in many customer-facing interactions. With Omnimics, many people will be able to predict and understand customer behavior based on tone, social media posts, and various disparate data that can now be connected.

Each of these AI types contributes to the rapidly evolving field of artificial intelligence, offering diverse capabilities that are reshaping industries and our interaction with technology.

Levels of AI by Capabilities

Additionally, you can choose to look at artificial intelligence through a different lens and that is of what the technology will be used for in the real world. Below are three classifications used frequently with examples.

- Artificial Narrow Intelligence (ANI) – AI specialized in specific tasks, such as self-driving cars, recommendation algorithms, and virtual assistants like Siri or Alexa.

- Artificial General Intelligence (AGI) - Hypothetical AI capable of performing any intellectual task a human can.
- Artificial Super Intelligence (ASI) - A theoretical form surpassing human intelligence in all domains.

Some of these technologies frighten people. Therefore, it is crucial to be specific about your objectives, how you plan to implement these technologies, and the guidelines for their use. I hate to keep emphasizing governance, but I believe that governance consultants must scale alongside the companies adopting these new innovations for everyone to get an optimal outcome of this process.

QUANTUM COMPUTING

While today's technology continues to improve, we are on the cusp of several dramatic leaps in the next decade. Nvidia's Blackwell chipset is a prime example—its advancements have made Nvidia a Wall Street favorite since OpenAI introduced ChatGPT in November 2022.
In recent months, the chatter around Quantum Computing has risen dramatically.

In my humble opinion, Quantum Computing is becoming a necessity. The amount of data and analytics requires more and more computing power. Chips and Processors are expanding exponentially at the same time, driven by demand for AI solutions across the board. This will likely impact the investments and results that may impact the trajectory of the quantum revolution.

Quantum computing and the companies driving innovation are poised for significant growth. A few of the companies I am tracking include Rigetti, Microsoft, and Google. Many other companies (including IBM) are forging strategic partnerships and technological advancements. Here's a breakdown of recent developments:

Rigetti Computing: Scaling Infrastructure and Partnerships Strategic Collaboration: Partnered with Quanta Computer in a $200+ million joint investment over five years to develop superconducting quantum systems, leveraging Quanta's server manufacturing expertise.

Technology Milestones:

- Launched the 84-qubit **Ankaa-3 system** with 99% median gate fidelity and plans for 100+ qubits by late 2025.
- Pioneered **AI-assisted calibration** and optical readout techniques for qubits, enabling faster scaling and reduced error rates.
- Sold its first quantum processing unit (QPU) to Montana State University, expanding academic R&D access.

Roadmap: Targets commercialization in 4–5 years, focusing on achieving 1,000+ qubits with 99.7% fidelity and real-time error correction.

Microsoft: Novel Qubit Design

Topological Qubits: Developed the **Majorana 1 chip**, leveraging a newly engineered topological superconductor state of matter. This approach aims to reduce volatility and simplify error correction

Timeline: Claims a fault-tolerant quantum computer could be operational within five years, with an eight-qubit prototype already submitted to DARPA.

Advantage: Combines semiconductor and superconductor technologies, potentially enabling million-qubit systems in compact designs.

Google: Accelerating Practical Applications

Breakthrough Timeline: Google Quantum A.I. predicts quantum applications surpassing classical computers within five years, targeting physics simulations and novel data generation.

December 2024 Milestone: Demonstrated quantum supremacy with a task completed in minutes that would take classical supercomputers a millennia to complete.

Overall Market Trajectory

Investment Surge: Rigetti's $35 million equity infusion from Quanta and Microsoft's DARPA-backed research highlight growing financial commitments.

Market Forecast: Superconducting qubit systems (favored by Rigetti and Microsoft) are projected to dominate due to scalability and semiconductor compatibility, with the quantum market expected to reach $1–2 billion annually by 2030.

Key Drivers of Growth

Error Rate Reduction: Rigetti's 2x error reduction target and Microsoft's stable qubit design address critical barriers to reliability.

Hybrid Architectures: Integration of AI (Rigetti) and classical computing infrastructure (Microsoft) enhances usability.

Industry-Academia Collaboration: Partnerships like Rigetti-Montana State and government programs (DARPA) accelerate talent development and R&D.

While timelines vary—Rigetti and Microsoft target 4–5 years for commercialization, Google aligns similarly— the convergence of scalable hardware, novel physics, and cross-sector collaboration suggests that quantum computing is transitioning from lab curiosity to near-term industrial relevance.

In the meantime, there is increased interest in the work undertaken by Oak Ridge National Laboratories (ORNL).

Oak Ridge National Laboratory (ORNL) has made significant strides in quantum computing development, positioning itself as a key player. Here's an overview of ORNL's recent quantum computing initiatives:
Strategic Collaborations

ORNL has formed partnerships with leading quantum computing companies to advance its quantum capabilities:

Quantum Brilliance Collaboration: ORNL works with Quantum Brilliance to integrate room-temperature diamond quantum accelerators into its high-performance computing (HPC) systems. This collaboration aims to:

- Explore parallel and hybrid quantum computing
- Develop new computational methods and software tools
- Inform the design of future HPC infrastructure

IonQ Partnership: ORNL and IonQ have developed a novel hybrid quantum algorithm based on the Quantum Imaginary Time Evolution (QITE) principle. This approach:

- Reduces the number of two-qubit gates needed by over 85%
- Outperforms other quantum optimization algorithms like QAOA
- Enables solutions for complex optimization problems on commercially available hardware

Quantum Roadmap and Research Focus: ORNL unveiled its Quantum Roadmap at the "Quantum on the Quad" event, showcasing its path toward quantum research and development. The laboratory's quantum initiatives include:

- Leveraging over 30 full-time quantum specialists and the Quantum Science Center
- Exploring practical applications of quantum mechanics in various sectors
- Focusing on four primary capabilities: quantum computing, quantum materials, quantum networking, and quantum sensing.

Integration with Classical Computing

- ORNL is working towards integrating classical and quantum computing systems:
- The Advanced Computing Ecosystem Testbed (ACE) serves as a centralized sandbox for deploying diverse computing resources.
- ORNL aims to connect quantum systems to its Frontier supercomputer, currently considered the world's most powerful classical supercomputer.

Quantum Computing User Program (QCUP)

ORNL has established the Quantum Computing User Program to provide researchers access to various commercial quantum computing resources7. This program includes:

- Access to IQM's Resonance quantum cloud service, featuring Crystal and Star topologies.
- Opportunities for researchers to explore new frontiers of quantum research and tackle complex scientific questions.

By leveraging its extensive resources, collaborations, and expertise, ORNL is playing a crucial role in advancing quantum computing technology and its applications in scientific research and critical U.S. infrastructure improvements.

WHAT DOES SUCCESS LOOK LIKE FOR YOU?

For executives, defining success isn't just about setting targets—it's about measuring the right things and asking the right questions. Success must be clearly articulated so teams understand not only the goals but also the why behind them.

The Power of Asking More Questions

The best leaders keep asking questions until they uncover the full picture. This relentless curiosity is what I call Omnimics—the pursuit of knowing everything that matters.

- Asking more questions ensures clarity.
- It signals to your team that you care about the right details.
- It aligns everyone toward a shared understanding of success.

Great executives leverage questions to cut through the noise, challenge assumptions, and refine their vision. The more questions you ask, the better you define what truly matters.

Executive Focus: The 'Why' Behind Metrics

Understanding the why behind every metric is critical. Too often, companies measure what's easy instead of what's meaningful. Defining actionable, relevant metrics is an iterative process—one that requires:

- **Multiple perspectives** to refine what's truly important.
- **Alignment across stakeholders** to ensure the organization is rowing in the same direction.
- **Adaptability** to update success metrics as goals evolve.

The best executives don't just dictate what to measure; they engage their teams in defining success together.

Conclusion

There is much to be excited about—and much to be cautious of—when it comes to emerging technologies. Every major innovation has followed a familiar pattern: initial fear, rapid adoption, and then the challenge of governance. AI, analytics, and quantum computing are no different. The real question isn't whether these technologies will be integrated into our world, but how we will guide their evolution in a way that benefits both businesses and society.

In my last company, we had two sayings: #GSD and Be Curious. In a world of Omnimics, these two value statements truly resonate with AI, analytics, and quantum computing. It is important to take the first step or the 500th in positioning yourself or your company to leverage this way of thinking. So, get things done. Create and deploy often. Whether it's code for an application or new processes, don't hesitate to involve a customer advisory board that can observe the new and exciting ways you and your company are innovating.

While you iterate on the use of these technologies, it is essential to remain curious. You can easily label this innovation. Consider what you are truly observing in response to your questions. Insights are always richer the second and third time around. I call this the Power of Three in analytics. Often, you must present for the first time to dispel myths and previous models for understanding the problem. During the second presentation, you begin to gain acceptance of key points in the value proposition. Customers usually validate these assumptions with employees and their inner circle in the community. By the third presentation, you gain acceptance by demonstrating that you heard the question, understood the answer, and have a recommendation on how to act on these insights.

You have achieved Omnimics.

RESOURCES

Competitive Intelligence: "I work for Caterpillar; how are my competitors using AI today?"

https://www.perplexity.ai/search/i-work-for-caterpillar-what-ar-pNHIGAavRKmkiGFu6.XOrw

Comparison Table: AI Initiatives by Major Caterpillar Competitors

Company	Key AI Applications	Notable Technologies/Initiatives
Komatsu	AI-powered site visualization, operator alerts, automation	SmartConstruction, NVIDIA AI partnership
Volvo CE	Operator-assist, compaction analysis, safety systems	Volvo Co-Pilot, Compact Assist, computer vision
John Deere	Precision agriculture, autonomous vehicles	See & Spray, full farm automation by 2030
Hitachi	Predictive maintenance, remote monitoring	ConSite® Mine, IoT & AI analytics
CNH Industrial	AI customer/dealer support, diagnostics	CNH AI Tech Assistant chatbot
SANY	Autonomous trucks, logistics automation	Pony.ai partnership, Level 4 autonomy
Terex	AI-powered recycling automation	ZenRobotics collaboration

Key Trends

- **Automation and Autonomy:** Many competitors are working toward fully or partially autonomous machinery for construction, mining, and agriculture.
- **Predictive Maintenance:** AI is being used to predict equipment failures and schedule maintenance before breakdowns occur, reducing downtime.
- **Operator Assistance and Safety:** AI-driven systems are enhancing operator awareness, providing real-time feedback, and preventing accidents.
- **Customer Support:** AI chatbots and virtual assistants are improving technical support and repair processes for dealers and customers.
- **Smart, Connected Machines:** Cloud connectivity enables continuous data collection and remote AI-based analytics for performance optimization.

Caterpillar's competitors are rapidly adopting AI to improve productivity, safety, and customer satisfaction, positioning themselves for a future of smarter, more autonomous heavy equipment 6 7 8 9 10 11 12 .

How Caterpillar's Competitors Are Using AI Today

Caterpillar's main competitors—including Komatsu, Volvo Construction Equipment, John Deere, Hitachi Construction Machinery, CNH Industrial, SANY, and Terex—are leveraging artificial intelligence (AI) to drive innovation, efficiency, and automation across the heavy equipment and construction industries. Here's a breakdown of how these companies are deploying AI:

Komatsu

- Komatsu is actively integrating AI into heavy equipment through partnerships, such as with NVIDIA, to deploy intelligent cameras and sensors on construction sites. Their SmartConstruction initiative uses AI-powered cameras, drones, and 3D laser scanners to create real-time visualizations of job sites, improve safety, and optimize equipment usage. The long-term goal is to enable autonomous machine operation and boost productivity by providing operators with real-time data and alerts .

Volvo Construction Equipment

- Volvo CE has developed AI-driven operator-assist technologies like the Volvo Co-Pilot system, which provides intelligent services (Load Assist, Dig Assist, Compact Assist, Pave Assist) via a tablet interface. Their Compact Assist uses neural networks for real-time compaction analysis, helping prevent over-compaction and improving quality. Volvo is also working on AI-based computer vision for object detection and operator safety, as well as connected machines for cloud-based AI solutions .

John Deere

- John Deere is investing heavily in AI, particularly in computer vision and machine learning, to support autonomous farming. Their See & Spray technology uses AI to detect and target weeds, reducing herbicide use and improving crop health. Deere's vision is a fully autonomous production cycle for major crops by 2030, using AI, robotics, sensors, and connectivity to automate planting, spraying, and harvesting .

Hitachi Construction Machinery

- Hitachi's ConSite® Mine platform leverages IoT and AI to remotely monitor mining equipment 24/7, analyzing operational data to predict maintenance needs and prevent costly failures. Machine learning algorithms are used to identify issues like cracks in excavator booms or arms, providing predictive alerts and detailed diagnostics to operators and dealers .

CNH Industrial

- CNH Industrial has launched the CNH AI Tech Assistant, an AI-powered chatbot that provides instant technical support and repair plans to dealer technicians. This tool is designed to reduce repair times, improve machine uptime, and enhance customer satisfaction by delivering fast, accurate diagnostics for CNH brand equipment .

SANY

- SANY, in partnership with Pony.ai, is developing Level 4 autonomous trucks for logistics and construction. These vehicles will use advanced AI for autonomous driving, integrating Pony.ai's virtual driver technology with SANY's expertise in heavy equipment manufacturing. The aim is to mass-produce intelligent, self-driving trucks for the global market .

Terex

- Terex is working with ZenRobotics to implement AI-driven robotic systems for recycling applications, automating the sorting and processing of construction and demolition waste to improve efficiency and sustainability .

Technologist Tools

Nvidia's Explore: https://build.nvidia.com/

AI Warehouse: https://www.thewarehouse.ai/

Microsoft CoPilot Tools/Devs/Azure Foundry: developer.microsoft.com

Google AI Studio Tools: https://aistudio.google.com/

Federal Government's Response to AI Policy

https://www.speaker.gov/wp-content/uploads/2024/12/AI-Task-Force-Report-FINAL.pdf

This report is a comprehensive overview of AI policy considerations for the US government. It emphasizes the immense potential of AI to improve lives and the need to proactively address potential risks. The report provides a framework for Congress to approach AI policy in a way that encourages innovation while safeguarding against harm.

Key Takeaways:

AI is a Big Deal: AI is recognized as a transformative technology that can reshape society and the economy. The US must maintain its leadership position.

Balance is Key: The core message is balancing AI innovation with responsible safeguards.

Guiding Principles: The report establishes key principles to guide AI policy:

Identify AI Issue Novelty: Determine if AI-related issues are genuinely new or already addressed by existing laws.

Promote AI Innovation: Encourage the development and deployment of AI technologies.

Protect Against AI Risks and Harms: Mitigate potential risks through technical and policy solutions.

Empower Government with AI: Use AI to improve government services and efficiency.

Support Sector-Specific Policies: Allow regulators to address AI within their areas of expertise.

Take an Incremental Approach: Adapt policies as AI evolves.

Keep Humans at the Center of AI Policy: Focus on the human impact of AI.

Specific Areas of Focus: The report delves into various critical areas:

Government Use: Calls for responsible AI implementation within federal agencies, focusing on transparency and workforce development.

Data Privacy: Highlights the need for robust data privacy policies to maintain consumer trust.

National Security: Emphasizes AI's importance in defense and the need to counter adversaries' AI capabilities.

Research & Development: Stresses continued investment

in AI research, standardization, and infrastructure.
Civil Rights & Liberties: Addresses potential biases and discrimination in AI systems.

Other Key Areas: Education and Workforce, Intellectual Property, Content Authenticity, Open and Closed Systems, Energy Usage and Data Centers, Small Business, Agriculture, Healthcare, and Financial Services.

Recommendations:

The report offers numerous recommendations across these areas, including:
Supporting flexible governance models for AI.
Promoting data access while protecting privacy.
Expanding AI training within the Department of Defense.
Encouraging public-private partnerships for AI research.

In Summary:

This report serves as a starting point for Congress to navigate the complexities of AI policy. It encourages a balanced approach that fosters innovation while protecting against risks, focusing on sector-specific regulations and continuous adaptation to the evolving AI landscape. For you as an executive, it highlights the key areas where AI will have a significant impact. It provides a framework for understanding the policy considerations that will shape its development and deployment.

Source: Perplexity March 11, 2025

The Omnimics Challenge: Mastering Thought and Learning in the Age of Infinite Information

For the next generation of IT leaders:

You stand at the precipice of a new era in technology leadership—one that I call the "Omnimics Age." This challenge is designed to prepare you for a future where information scarcity is a relic of the past, and your success hinges on your ability to think critically and learn rapidly.

The Omnimics Advantage

Imagine having access to every piece of information in any format. This is the reality you'll soon face. Unlike previous generations who grappled with limited data and incompatible systems, you'll have the world's knowledge at your fingertips. But with great power comes great responsibility—and a new set of challenges.

Your Mission: Think and Learn

In this new landscape, your value as an IT leader will be defined by two core competencies:

1. Thinking:

Your ability to analyze, synthesize, and strategize will be paramount. You must:

- Dissect complex problems with precision
- Envision innovative solutions that others might overlook
- Navigate the deluge of information to identify what truly matters
- Learning:

With the rapid pace of technological change, your capacity to acquire and apply new knowledge will be crucial. You must:

- Absorb vast amounts of information quickly and effectively
- Adapt to new technologies and methodologies at lightning speed
- Cultivate a mindset of perpetual growth and curiosity

The Challenge

Here's how you'll prove your mettle in the Omnimics Age:

1. Problem Presentation: You'll be given a complex IT challenge that requires a multifaceted solution.
2. Strategic Thinking: Analyze the problem from all angles. Consider the broader implications for your organization and stakeholders.
3. Consensus Building: Engage with key stakeholders to define the desired outcome. Your ability to communicate and align diverse perspectives will be tested.
4. Rapid Learning and Execution: Leverage your Omnimics resources to gather all relevant information. Learn what you need to know—fast. Then, implement your solution with precision and agility.
5. Reflection and Iteration: After execution, assess your process and results. What did you learn? How can you improve for the next challenge?

Your Edge in the Omnimics Age

Remember, in a world where everyone has access to everything, your competitive advantage lies not in what you know but in how you think and how quickly you can learn and apply new knowledge.

Hospital Management: A Metrics-Driven Approach

Case Study: Implementing Omnimics in Healthcare Financial Analysis

Executive Summary

This case study examines how hospital executives can leverage comprehensive metrics analysis ("Omnimics") to enhance decision-making and operational efficiency. Focusing on financial, operational, and clinical metrics, the study demonstrates how granular data analysis can transform hospital management and sustainability.

Background

Modern hospital CEOs face complex challenges that require data-driven decision-making across multiple domains:

- Financial Metrics: Measuring economic stability and performance
- Operational Metrics: Tracking efficiency and process effectiveness
- Quality Metrics: Ensuring high standards in patient care
- Patient Satisfaction: Gauging patient experiences

The healthcare industry's complexity demands sophisticated analytics to maintain the delicate balance between mission-driven care and financial sustainability, as captured in the expression: "No money, no mission."

Challenge

A mid-sized hospital needed to comprehensively understand its financial performance at the service line level while integrating this data with operational and clinical metrics.

Leadership required actionable insights to:

- Identify unprofitable service lines
- Understand payer mix impact on revenue
- Optimize Billing Practices/Revenue Cycle Management
- Create sustainable financial models while maintaining quality care

Approach: Implement Omnimics Framework

Phase 1: Financial Metrics Implementation

1. Patient-Level Financial Classification

The hospital established a granular financial classification system with multiple layers:

a. Financial Class (Broad Categories)

- Commercially insured
- Medicare
- Medicaid
- Federal payers
- Self-pay patients

b. Payer Detail (Sub-Categories)

- Medicare vs. Medicare Advantage vs. Medicare Supplement Plans
- Commercial insurance vs. self-insured employer plans (where insurers act as third-party administrators)
- Medicaid variations
- Uninsured (resulting in "no pay" write-offs)

2. Charge and Payment Analysis

The hospital implemented a sophisticated system to track:

- **Charge Master Management:** Detailed contractual agreements between payers and the provider
- **Procedure-Based Pricing Analysis:** Example of total knee joint replacement (CPT code 27447)
- Medicare allowed charges: ~$25,000
- Commercial insurance charges: $75,000-$80,000
- Negotiated rates: ~$30,000 for commercial insurance
- **Additional Financial Considerations:**
  - Patient deductibles and co-pays
  - Self-pay negotiated rates under price transparency regulations
  - Specialized employer-provided arrangements (e.g., Walmart's spine discount with Mayo Clinic)
  - Additional service charges (bed fees, specialist consultations, surgical suite fees)

3. Data Infrastructure Development

The hospital implemented the following:

- **EMR Integration:** Leveraging Epic (Cerner, Meditech, or MedHost) for detailed cost accounting
- **Enterprise Data Warehouse:** Enabling on-demand queries at the individual patient level
- **Line-Item Detail Database:** Creating patient-specific financial profiles for analysis
-

Phase 2: Customer Analytics Integration

1. Business Customer Profiling

- For B2B services, the hospital developed a business personality profiling system to categorize other businesses as:
- Innovators
- Early adopters
- Laggards
- Technology-resistant ("Luddites")

2. Consumer Demographics Analysis

The hospital collected and analyzed:

- Age
- Gender
- Race
- Household income
- Educational attainment

3. Geographic Analysis

Implementation of location-based analytics:

- Address-level precision with latitude/longitude coordinates
- Hierarchical geographic groupings:
- Block
- Block Group
- Census tract
- County
- State
- Core Business Statistical Area (CBSA)

4. Psychographic Profiling

The hospital utilized advanced consumer behavior analysis:

The facility implemented ESRI Tapestry, a leading psychographic profiling system (Alternatively you could use Prism by Claritas)

The analysis included a selection of 2,080+ variables related to consumer spending habits and preferences. Additional data aggregation was used from multiple sources (surveys, loyalty cards, credit card data).

The final step included the identification and selection from 68 unique household types to predict service line utilization, profitability, patient procedure predictions, and social determinants of health to create an impact analysis on the facility, the community, and the patient.

Results

The implementation of the Omnimics framework allowed the hospital to:

- Calculate precise profitability metrics per patient and service line
- Make data-driven decisions about service offerings
- Optimize payer contracts based on actual performance data
- Predict which consumer segments would utilize specific services
- Develop marketing strategies targeted to specific demographic and psychographic profiles
- Ensure long-term financial sustainability while maintaining mission focus

Key Learnings/Takeaways

- **Integrated Data is Essential:** Siloed financial, operational, and clinical data prevents comprehensive analysis.
- **Granularity Matters:** Patient-level financial analysis provides insights impossible to obtain from aggregate data.
- **Beyond Demographics:** Psychographic and geographic analysis provides deeper insights into consumer behavior than demographics alone.
- **Financial Complexity:** Understanding the nuances of healthcare billing, including charge master details and payer agreements (price transparency), is crucial for accurate financial analysis.

- **Technology Infrastructure:** EMR systems and enterprise data warehouses are foundational requirements for implementing Omnimics.

Conclusion

The hospital's adoption of a comprehensive Omnimics approach transformed its financial analysis capabilities. The organization achieved better financial outcomes by integrating detailed financial metrics with customer analytics while maintaining its commitment to quality patient care.

This case study demonstrates that healthcare organizations can successfully navigate the complex balance between mission and margin with the right data infrastructure and analytical framework.

Next Steps and Recommendations

- Expand the Omnimics framework to include more detailed clinical outcome metrics.
- Develop predictive models for patient utilization based on psychographic profiles.
- Implement real-time financial dashboards for service line managers
- Create training programs to help staff understand and utilize financial metrics
- Regularly review and update charge master based on profitability analysis

Retail Banking: Optimizing Customer Value Through Metrics Analysis

Case Study: Implementing Omnimics for Checking and Savings Accounts

Executive Summary

This case study examines how a mid-sized regional bank leveraged comprehensive metrics analysis ("Omnimics") to enhance its retail banking operations, specifically focusing on checking and savings accounts. By implementing granular data analytics, the bank transformed its approach to customer profitability, product offerings, and marketing strategies, significantly improving customer retention and account profitability.

Background

Modern retail banks face increasing competition from traditional competitors, digital-only banks, and fintech companies. To remain competitive, banks must develop a deep understanding of:

- **Customer Profitability:** Measuring the net contribution of each customer
- **Product Performance:** Evaluating the performance of various financial instruments
- **Customer Behavior:** Understanding usage patterns and preferences
- **Market Positioning:** Differentiating offerings in a crowded marketplace

The complexity of today's banking environment requires sophisticated analytics to balance growth, profitability, and customer satisfaction.

Challenge

A regional bank with 75 branches across three states was experiencing declining profitability in its retail banking division, particularly in checking and savings accounts.

The executive team needed to:

- Identify which customer segments were most profitable
- Understand the true cost and value of different checking and savings products
- Develop targeted strategies to acquire and retain high-value customers
- Optimize fee structures and interest rates without driving away customers
- Create a more personalized approach to customer relationships
- Approach: Implementing Omnimics Framework

Phase 1: Financial Instrument Metrics Implementation

1. Account-Level Financial Classification

The bank established a detailed classification system for retail banking products:

Checking Account Categories

- Basic checking (no minimum balance)
- Premium checking (with minimum balance requirements)
- Interest-bearing checking

- Student Checking
- Senior Checking
- Business checking (small business)

Savings Account Categories
- Standard savings
- High-yield savings
- Money market accounts
- Certificates of deposit (various terms)
- Holiday/special purpose savings
- Youth/student savings

2. Revenue and Cost Analysis

The bank implemented a comprehensive system to track:

Revenue Components:
- Interest margin (difference between deposit rates and lending rates)
- Monthly maintenance fees
- Transaction fees (overdraft, wire transfers, etc.)
- Interchange revenue from debit card usage
- Cross-sell revenue from related products

Cost Components:
- Account servicing costs
- Transaction processing costs
- Branch operating expenses (allocated)
- Digital banking costs

- Customer service expenses
- Marketing acquisition costs
- Regulatory compliance costs

3. Customer Profitability Metrics

The bank developed metrics to assess each customer's contribution:

- **Customer Lifetime Value (CLV):** Projected net profit from customer over relationship lifespan. Each CLV was created by the initial account type the customer began the relationship with the banking institution.
- **Relationship Depth:** Number of products per customer
- **Account Activity Metrics:** Transaction volumes, deposit balances, digital engagement
- **Cost-to-Serve:** Personalized assessment of servicing costs based on channel preferences
- **Attrition Risk Score:** Predictive model for likelihood of account closure

4. Data Infrastructure Development

The bank implemented:

- **Core Banking System Integration:** Capturing detailed transaction data
- **Customer Data Platform:** Creating unified customer profiles
- **Data Warehouse:** Enabling complex queries across product lines
- **Real-time Analytics Dashboard:** Providing on-demand profitability analysis

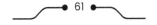

Phase 2: Customer Analytics Integration

1. Demographic Analysis

The bank collected and analyzed:

- Age
- Gender
- Income level
- Occupation
- Family life stage
- Educational attainment

2. Geographic Analysis

Implementation of location-based analytics:

- Branch proximity
- Home/work location mapping
- Address-level precision with census tract integration
- Neighborhood economic indicators
- Market penetration by geographic segment

3. Behavioral Analysis

The bank tracked:

- Channel preferences (branch, mobile, online, ATM)
- Transaction patterns (frequency, amounts, timing)
- Product usage behaviors
- Bill pays and direct deposit utilization
- Digital feature adoption

4. Psychographic Profiling

The bank utilized advanced consumer analysis:

- Implementation of a new financial personality model
- Risk tolerance assessment
- Financial goals and aspirations
- Spending and saving patterns
- Brand affinity and loyalty drivers
- Implementation of ESRI Tapestry segments

Results

The implementation of the Omnimics framework allowed the bank to:

Identify Profitability Drivers: The bank discovered that 22% of customers generated 78% of retail banking profits, while 35% of accounts were unprofitable.

Optimize Product Offerings: The bank redesigned its checking account lineup, reducing from eight to four options while increasing overall profitability by 14%.

Implemented Targeted Fee Strategies: Fee waivers were aligned with behaviors that increased profitability, resulting in a 23% reduction in fee complaints while maintaining fee income.

Enhanced Cross-Selling: Predictive models identified optimal timing for product recommendations, increasing cross-sell success rates by 31%.

Reduce Attrition: Customer retention improved by 18% through proactive engagement strategies with at-risk accounts.

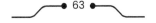

Personalized Marketing: Campaigns targeted to specific psychographic segments saw 42% higher response rates than generic promotions.

Key Learnings/Takeaways

Holistic Profitability View: Account-level metrics alone are insufficient; customer-level profitability analysis revealed relationships between product usage and overall value.

Behavior Matters More Than Demographics: Traditional demographic segmentation proved less predictive of profitability than behavioral and psychographic factors.

Digital Engagement Correlation: Customers with high digital engagement had 2.7x higher profitability and significantly lower attrition rates.

Fee Sensitivity Varies: Psychographic segments showed dramatically different sensitivity to fees, allowing for targeted fee strategies.

Life Stage Transitions: Major life events (marriage, children, retirement) represented critical intervention points for deepening banking relationships.

Infrastructure Investment Payoff: The substantial investment in data infrastructure generated ROI within the first 12 months through improved decision-making.

Conclusion

The bank's adoption of a comprehensive Omnimics approach transformed its understanding of retail banking profitability. By integrating detailed financial metrics with customer analytics, the organization achieved significant improvements in profitability while enhancing customer satisfaction through more personalized approaches.

This case study demonstrates that retail banks can successfully navigate the complex balance between fee income, interest margins, customer experience, and long-term relationship value with the right analytics framework.

Next Steps and Recommendations

- **Expand Beyond Deposit Products:** Apply Omnimics approach to lending products and wealth management.

- **Real-time Personalization:** Implement AI-driven personalization in digital banking platforms

- **Predictive Attrition Models:** Enhance early warning systems for at-risk relationships

- **Branch Network Optimization:** Use geographic and profitability data to optimize the branch network. Opening of new branches in underserved geographies. Moved branches to Customer Pods based on geospatial analysis and traffic flows.

- **Employee Performance Integration:** Align staff incentives with customer profitability metrics

- **Advanced Financial Wellness Tools:** Develop digital tools that promote customer financial health while deepening relationships

- **Alternative Data Integration:** Incorporated non-traditional data sources for enhanced customer insights

MAJOR TECHNOLOGY MILESTONES

Early Internet Development

1969: ARPANET established, connecting four major universities in the southwestern US

1971: Ray Tomlinson sends the first email (using the @ symbol to separate the user from the host)

1973: ARPANET makes its first international connections to England and Norway

1983: ARPANET standardizes on TCP/IP protocol

World Wide Web

1989: Tim Berners-Lee proposes the World Wide Web at CERN

1991: First web page created by Tim Berners-Lee

1993: CERN announces that the World Wide Web will be free to everyone

1994: First World Wide Web conference held at CERN

1995: Commercialization of the Internet begins in earnest as NSF ends its sponsorship of the Internet backbone

Video Streaming Technology

1995: Microsoft NetShow released

1997: RealNetworks' RealPlayer achieves widespread adoption

2005: YouTube launched

2007: Netflix begins streaming service

2008: Hulu launched

2010: Vimeo introduces HD video capability

Social Media

1997: Six Degrees, one of the first recognizable social media sites, launched

2002: Friendster launched

2003: MySpace founded

2004: Facebook launched (initially just for Harvard students)

2006: Twitter founded

2007: Facebook opens to the public

2010: Instagram launched

2011: Snapchat founded

2016: TikTok launched (internationally)

Artificial Intelligence

1956: The term "artificial intelligence" coined at Dartmouth Conference

1997: IBM's Deep Blue defeats chess champion Garry Kasparov

1998: Google founded, revolutionizing search with PageRank algorithm

2011: IBM Watson defeats champions at Jeopardy!

2014: Amazon Alexa launched

2016: Google's AlphaGo defeats world champion Go player Lee Sedol

2018: Google demonstrates Duplex AI making natural phone calls

2022: ChatGPT released by OpenAI (November)

2023: Google Bard released (March)

Google Gemni released (December)

Anthropic's Claude AI assistant made publicly available (March)

Microsoft Copilot launched, integrating AI into Microsoft 365 (March)

2024:

Significant advancements in multimodal AI models that can understand and generate text, images, and code

Quantum Computing

1980: Paul Benioff describes the first quantum mechanical model of a computer

1994: Peter Shor develops a quantum algorithm for factoring large numbers

1998: The first working 2-qubit quantum computer demonstrated

2011: D-Wave Systems announces the first commercially available quantum computer

2016: IBM makes a quantum computer available to the public via the cloud

2019: Google claims quantum supremacy with its 53-qubit Sycamore processor

2021: IBM unveils 127-qubit quantum processor "Eagle"

2023: IBM announces 433-qubit "Osprey" quantum processor

2024: Several companies demonstrate practical quantum error correction techniques

2025: Microsoft's Topological Computing Platform

Google's Sycamore Computing Platform

ORNL Advancements in quantum materials and processing

Rigetti's Superconducting Processors

IBM's Quantum Cloud Services

IONQ Trapped-Ion Architecture

FOOTNOTES:

https://en.wikipedia.org/wiki/History_of_the_web_browser

Russell Ackoff, a systems theorist, is widely credited with formalizing and presenting the DIKW (Data-Information-Knowledge-Wisdom) hierarchy in his 1989 article "From Data to Wisdom".

(Lance Hawley) https://news.harding.edu/2025/01/the-wisdom-of-asking-the-right-questions.html

NOTES: